Backpack Bear's

Mammal Book

Written by Alice O. Shepard

Starfall®

Starfall Education, P.O. Box 359, Boulder, CO 80306

ISBN: 978-1-59577-086-8

The Animal Kingdom

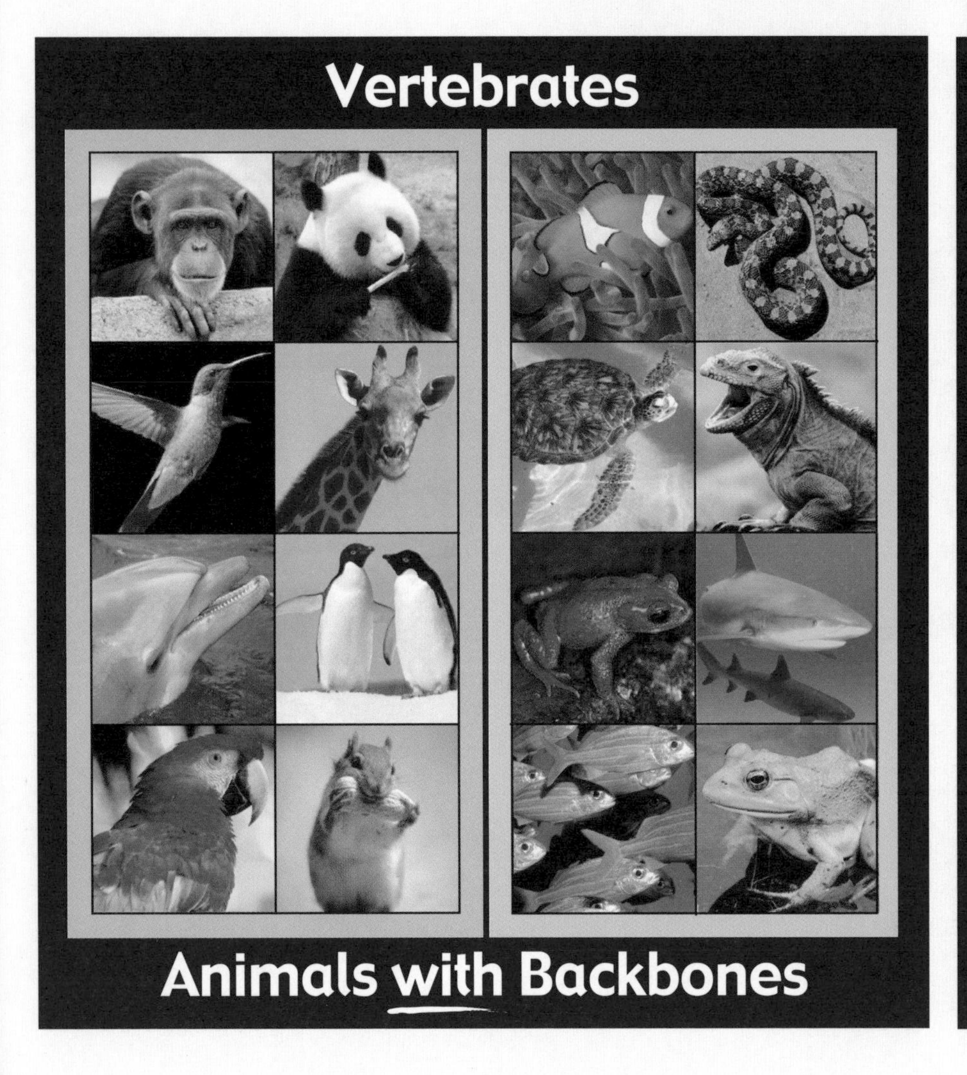

The Animal Kingdom can be divided into two groups: animals with backbones and animals without backbones.

Of the animals that have backbones, some are "warm-blooded" and some are "cold-blooded."

This book is about mammals.

Vertebrates (Animal

"Warm-Blooded"

with Backbones)

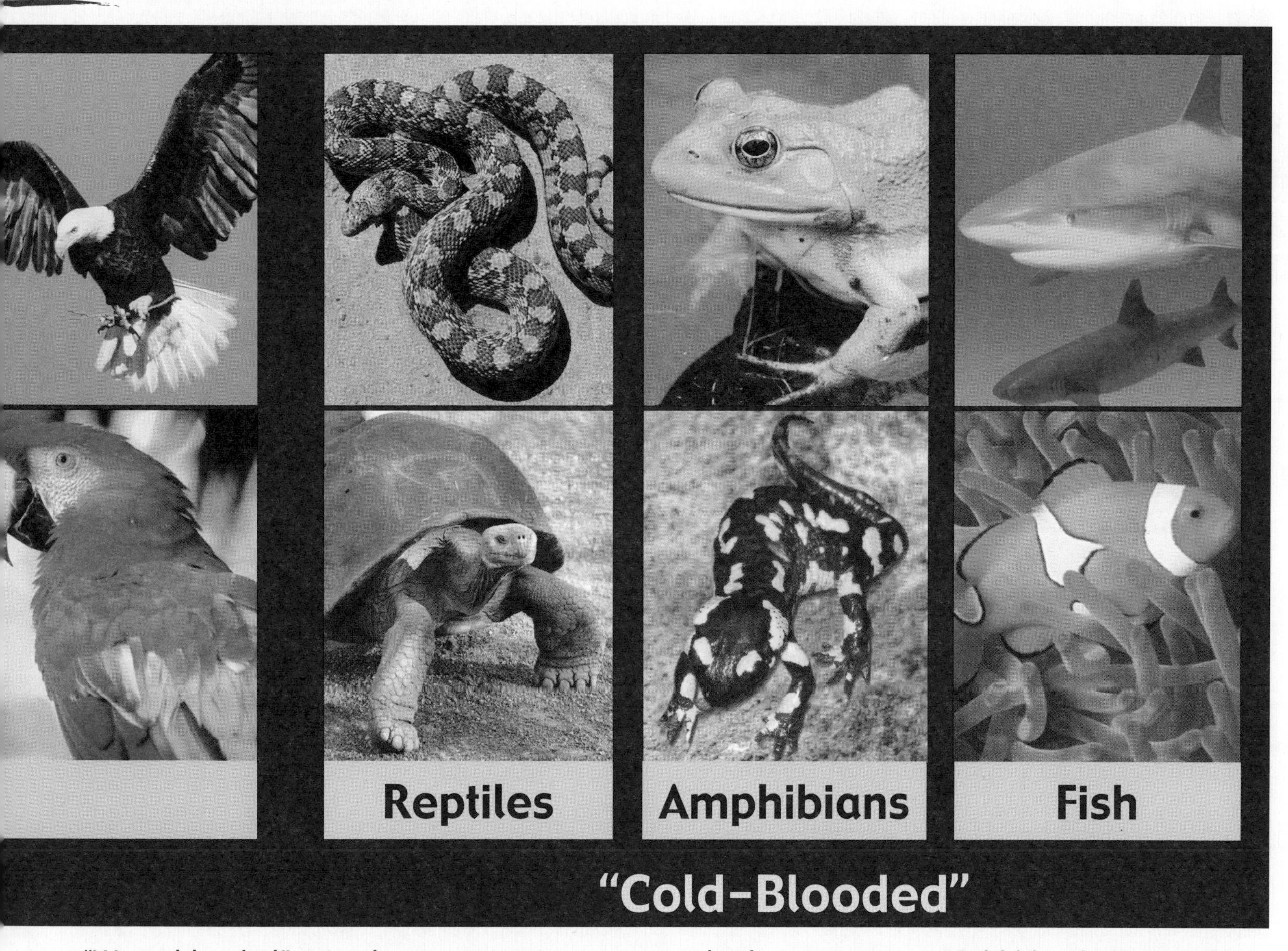

"Warm-blooded" animals can maintain a constant body temperature. "Cold-blooded" animals generally are not able to do this. Their body temperature changes depending on their surroundings.

Bengal Tigers

A ***mammal*** is a type of "warm-blooded" animal with a backbone. Two ***characteristics*** that make an animal a mammal are:

- Mammals feed their babies milk that comes from the mother's body.
- Mammals have hair, fur, or layers of fat to keep their bodies warm.

Most mammal babies begin their lives inside their mother's body. When they are born, most mammal babies look like their parents.

American Saddlebred Horses

Some mammals have one baby at a time. Others can have as many as five or more at one time!

Mammals look for safe places to have their babies.

Pigs

Baby kangaroos and koala bears are very tiny when they are born.

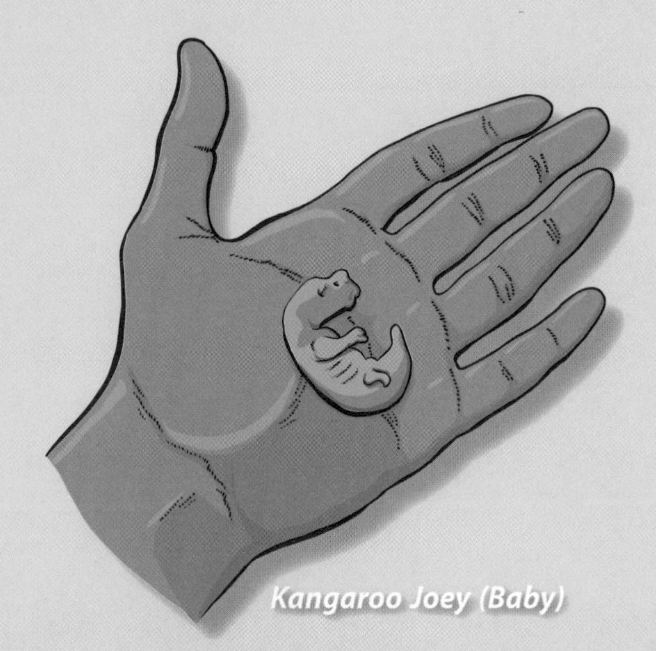
Kangaroo Joey (Baby)

They finish growing inside their mothers' pouches.

Koalas

Kangaroos

Platypus

Echidna

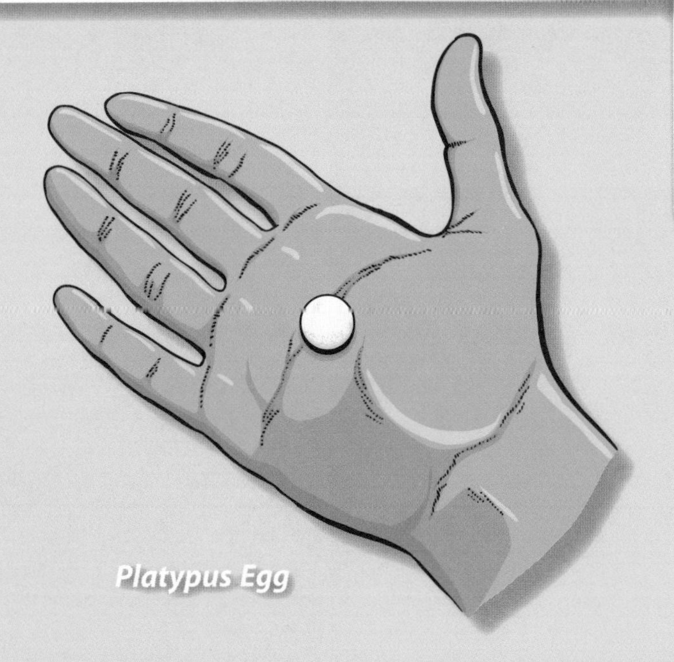

Platypus Egg

The babies from two very unusual mammals, the platypus and the echidna, hatch from eggs!

Some mammals are big!
Some are small.
Some mammals are tall,
others are short.

Mammals have four ***limbs***.
Most mammals use all
four of their limbs to
walk on land.

Highland Cattle

Giraffe

Cat

Spotted Fawn

White-Faced Monkey
FOREST FRIENDS
Wolf
Lambs
Lion
Human
Prairie Dog
Dog

Orangutan

Humans

Humans

Chimpanzees

Some mammals can walk on two of their limbs, and
use their other two limbs to hold things.

Do you know of any mammals that do this?

Not all mammals live on the land. Some mammals live underwater! They use their limbs to swim.

Whale

Mammals living under the water must come up to breathe air. This is because mammals breathe air with lungs.

I use a snorkel to go swimming.

Bats

One mammal, the bat, uses its limbs to fly through the air.

As mammals grow up, they stop drinking their mother's milk.

Some mammal parents teach their babies to find water and food so they can ***survive*** on their own.

Doe and Fawn

Bear and Cubs

Ewe and Lamb

Gorilla and Infant

Some mammals are ***predators***. They learn to survive by hunting and eating other animals.

Predators have characteristic sharp teeth and claws.

Leopard

Wolf

Other mammals are not predators.

They eat mostly plants.

Goat

Panda

Some mammals stay together in a group. They take turns watching for predators.

If they see a predator coming, they run away, fast!

Impalas

Other mammals protect themselves from predators by hiding.

Look at this mammal. The color of its fur helps it ***camouflage***, or blend in with its surroundings.

Fawn

Cow and Calf

Elephant and Calf

Horse and Foal

Seal and Pup

Mammals are "warm-blooded" animals with backbones.

They can be big or small. They can live on land, live underwater, or fly through the sky!

Two characteristics that make an animal a mammal are:

- They feed their babies milk that comes from the mother's body.
- They have hair, fur, or layers of fat to keep their bodies warm.

Glossary

Camouflage: An animal's natural coloring that helps it blend in with its surroundings

Characteristic: A quality or feature typical of something that helps us to identify it

Limbs: An arm, leg, or flipper of an animal

Mammal: A "warm-blooded" vertebrate that feeds its babies milk that comes from the mother's body

Predator: An animal that hunts and eats other animals to survive

Survive: To continue to live

What does it mean to be "Warm-Blooded"?

"Warm-blooded" animals can keep their body temperature nearly constant even though the temperature outside may change.

When the outside temperature becomes too cold, "warm-blooded" animals can keep their body warm. When the outside temperature becomes too hot, they can cool down. "Warm-blooded" animals can keep their body temperature constant in several ways.

When it's cold outside

"Warm-blooded" animals have hair, fur, feathers, or fat to help keep their body warm, but sometimes this isn't enough. Have you ever started to shiver when it was cold outside? Shivering is a way to make heat inside the body.

When it's hot outside

Some furry, "warm-blooded" animals, such as dogs, cool down by sticking out their tongues and breathing hard. This is called panting. Other animals, such as humans, sweat to cool down.

Keeping the body at a constant temperature requires a lot of energy. Animals get energy from the foods they eat. "Warm-blooded" animals must eat more food than "cold-blooded" animals to survive.

See *Backpack Bear's Reptiles, Amphibians & Fish Book* for information about "cold-blooded" animals.

Index

About the Author

Alice O. Shepard was born in Southampton, England. The "O" stands for Ophelia. Alice moved to the United States by boat when she was 15 years old. She was nervous about moving to a new home so far away and leaving her friends behind. Luckily, one of her friends came along—her cat Penelope! All during the cruise Alice was nervous about meeting new people. Penelope, on the other hand, was nervous about being surrounded by so much water!

Acknowledgements

Special thanks to Dr. Karen McBee, Department of Zoology, Oklahoma State University, for helping to check this book for accuracy. Thanks also to Aria Beemer.

Photo Credits

The following images were used by permission of these photographers from Flickr.com: *Echidna*, on page 11, © Max Sutcliffe; *Monarch butterfly*, on page 2, © Pieceoflace Photography. The following images were licensed from iStockphoto: *Koala*, on page 10, © Susan Flashman; *Bats*, on page 18, © Paul Cowan. *Whale*, on page 16, was used by permission from © James Watt, OceanStock.com.